NAME ....

ADDRESS..

PAGES	SUBJECT

PAGES	SUBJECT

BEND